Brain Exercise

For The Whole Family

Brain Teasers Riddles, Puzzles, Trivia Matching, And More To Keep Your Mind Young And Nimble.

Table Of Content

Copyright 2020 by Marcel Cohen All rights reserved

This document is geared towards providing exact and reliable information in regards to the topic and issue covered. The publication is sold with the idea that the publisher is not required to render an account, officially permitted, or otherwise, qualified services. If advice is necessary, legal or professional, a practiced individual in the profession should be ordered. -from a Declaration of Principles, which was accepted and approved, equally by a Committee of the American Bar Association and a Committee of Publishers and Associations.

In no way is it legal to reproduce, duplicate, or transmit any part of this document by either electronic means or printed format. 'the mottling of this publication is strictly prohibited, and any storage of this document is not allowed unless with written permission from the publisher. All rights reserved.

The information provided herein is stated to be truthful and consistent, in that any liability, regarding inattention or otherwise, from any use or abuse of any policies, processes, or directions contained within is the solitary and utter responsibility of the recipient reader. Under no circumstances will any legal responsibility or blame be held against the publisher for any reparation, damages, or monetary loss due to the information herein, either directly or indirectly. Respective authors own all copyrights not held by the publisher.

The information herein is offered for informational purposes solely and is universal as so. The presentation of the information is without a contract or any type of guarantee assurance.

The trademarks that are used are without any consent, and the publication of the trademark is without permission or backing by the trademark owner. All trademarks and brands within this book are for clarifying purposes only and are property of their owners, in any way affiliated with this document.

Mind benders 60 tricks

Table Of Content

TABLE OF CONTENT .... **IV**

MIND BENDERS **1**

MIND TICKLER **23**

MIND BREAKERS **31**

MIND BOGGLERS **44**

ANSWERS **57**

Blank Page

Mind Benders

Timber Land

A real estate agent wished to divide the piece of land shown below into four equal portions with each portion having two trees on it.

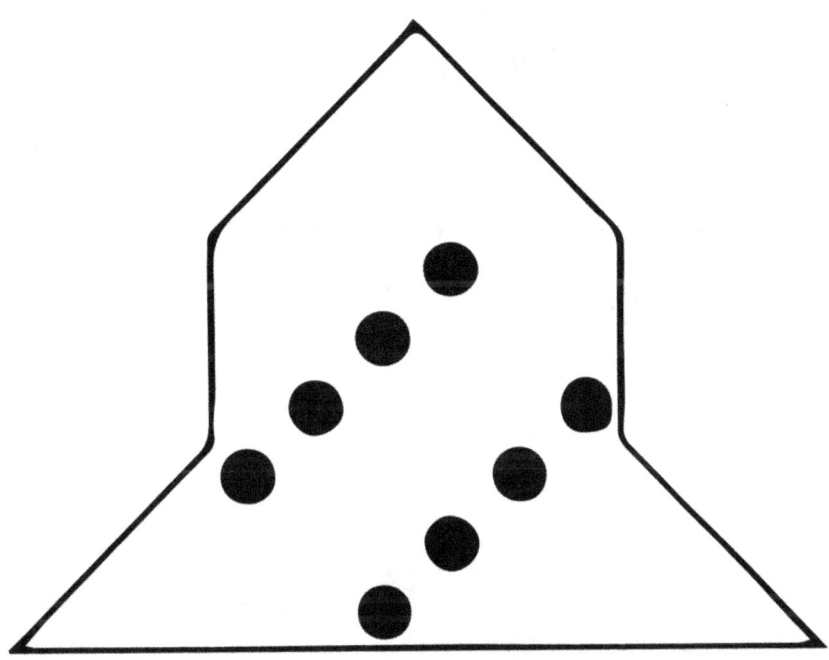

How would you recommend that he divide the land?

Rare Window

A young junior executive had just one window in his small office. The window measured four feet across from left to right and four feet from top to bottom.

The window was a perfect square.

Since the junior executive was a hard worker, he soon earned a promotion. Befitting his new position, he was given a more substantial office. This office also had just one window, which was four feet across, four feet high, and a perfect square. However, the window in the new office was twice the size (in the area) of the window in his old office. How was this possible?

Bus Stop

Melvin, who lives in Centreville, has one girlfriend living in North Centreville and another girlfriend living in South Centreville. He regards the two girls with equal favor and therefore decides to leave to chance the choice of which one he will visit each Sunday. Melvin knows that there is an equal number of buses running to each destination at regular ten-minute intervals. Thus he simply goes to his bus stop every Sunday without regard for the time of day and takes the first bus to arrive. Somehow it happens that on nine out of ten Sundays Melvin visits his South Centreville girlfriend. Why does this occur?

The Escalator

Each step of an escalator is 8 inches above the previous step, and the total vertical height of the escalator is 20 feet. The escalator moves up one-half step a second. If I step on the lowest step at the moment when it is level with the lower floor and walk-up at the rate of one step a second, how many steps do I take to reach the upper floor? (Do not include the steps taken to step on and off the escalator.) Fractured Fraction Twice a fraction plus half that Fraction, this sum multiplied by that Fraction, equals that Fraction. What is the Fraction?

Magic Square

In the illustration below the numbers, 1 through 9 are arranged consecutively in three rows of three numbers so as to form a "square."

$$\begin{matrix} 1 & 2 & 3 \\ 4 & 5 & 6 \\ 7 & 8 & 9 \end{matrix}$$

Can you rearrange the numbers within the square so that no matter which way you total three numbers inline
—vertically, horizontally, or diagonally—
will the sum always total 15?
When you find the correct arrangement,
you will have a "magic square."
 Can you think of a general rule for creating this kind
of a square with any nine numbers in succession?

Cannonball Express

Two railroad trains (on different tracks) are approaching one another on a straight railroad line, each train traveling at a speed of 30 mph. Suppose a fly— a very fast-flying fly— traveling at 60 mph, starts from the front of train A when the two trains are ten miles apart. It flies until it reaches the front of train B, turns around immediately and flies back to train A, where it turns around again and starts back to B. The fly continues to travel back and forth between the two trains, making shorter and shorter journeys, until the trains pass one another on the tracks. What is the total distance flown by the fly?

Tick-Tock Teaser

Suppose you were shipwrecked and marooned on a desert island with three watches, one who didn't run at all, one which you knew to be losing one minute a day, and one which was gaining one minute a day. Which watch would be most likely to show the correct time if you were to glance at the watches at any particular moment? Which watch would be least likely to show the correct time?

Pioneer Puzzle

A pioneer family, two children, and their parents started to trek cross-country to California. Setting out on foot, they managed to reach the Mississippi River. They were wondering how to cross it when they spotted a small row-boat. But they quickly realized that the boat would hold no more than one adult and one child — it could not safely carry both parents at the same time.
How did the entire family reach the other side of the river?*

Busy Worm

A worm climbs up a telephone pole six feet during the day, and slides back five feet during the night. If the telephone pole is 30 feet high and the worm starts from ground level, how many days does it take the worm to reach the top of the pole?

Lightning Multiplier

When the famous mathematician Karl Gauss was only seven years old; his teacher gave him and the rest of the class this problem: What is the sum of all the numbers from 1 to 100? To the teacher's surprise, Karl had the answer in a few seconds, while the rest of the class was still busily calculating. What was the answer, and how did he reach it so quickly?

Collective Farm

A farmer plans to divide his land equally among his four sons. But the farmer's land is L-shaped, as in the drawing below, and the farmer can't figure out how to split the land into four parts of equal area, each part having the same shape. Can you help him?

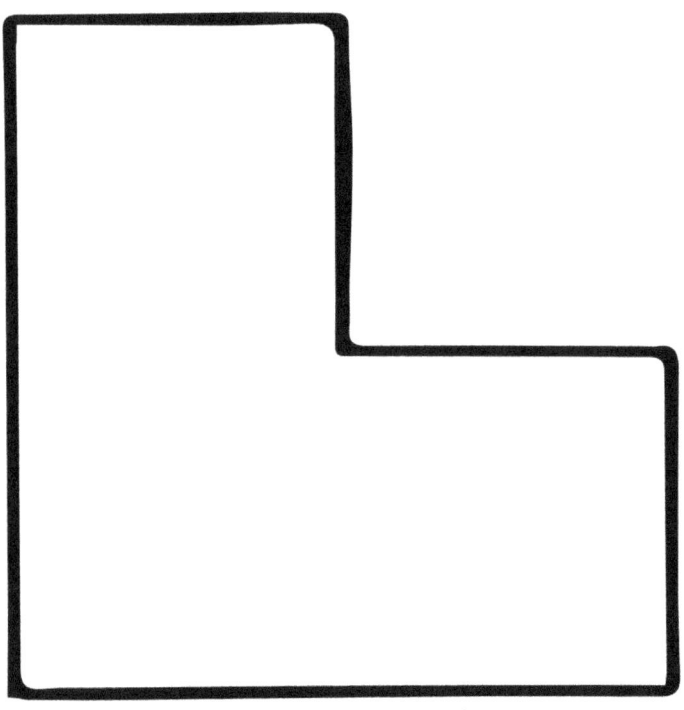

Happy Hunter

A game hunter, captured by hostile natives, was allowed to determine his own fate. He was ordered to make one definite statement, not a generalization.

Should the statement be false, the hunter would die by fire. Should it be true, he would die by poison.

With quick thinking, the hunter made a statement that made it impossible for the natives to kill him by either fire or poison and thus saved his life.

What did he say?

Does 1 = 2?

See if you can find the fallacy in this mathematical reasoning.

Let $a = b$

(1) $ab = b^2$ (by multiplying both sides by b)

(2) $ab - b^2 = a^2 - b^2$ (by subtracting b^2 from both sides)

(3) $b(a - b) = (a + b)(a - b)$ (by factoring both sides)

(4) $b = (a + b)$ (by dividing both sides by (a-b)

(5) $b = b + b$ or $b = 2b$ (by substituting b for a)

(6) $1 = 2$ (by dividing both sides by b)

Number Stumper

Place the numbers 1 through 10 in the circles of the figure below.

Arrange the numbers so that no two consecutive numbers are connected by any lines in the figure.

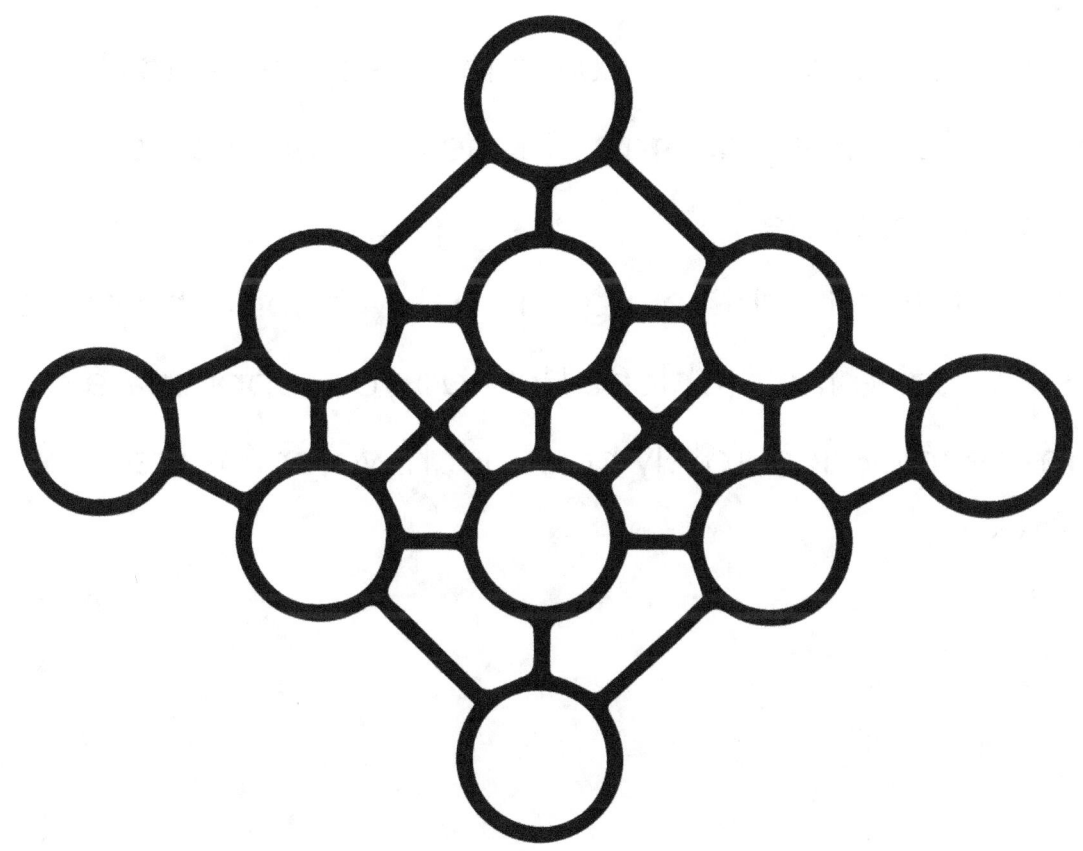

Wet Solution

Prior to a predicted drought, the inhabitants of a small town wanted to know how many gallons of water they had in their reservoir— a nearby lake.

It was safe to assume that during the next two weeks the change in the amount of water in the lake would be negligible.

Engineers had tried to measure the lake's area and determine its varying depths, but the measurements involved were too inaccurate.

It was left to a bright young chemist with some harmless red dye to solve the problem. How do you suppose he could determine fairly accurately how much water the reservoir contained?

Weighty Problem

Suppose you had eight marbles, all identical in appearance, and you knew one marble was heavier than the others. Using a balance scale, how could you be sure of finding the heavier marble in no more than three weighings?

Who's Who?

Imagine a small town in which the jobs of butcher, grocer, druggist, and policeman belong to four men named Clark, Jones, Smith, and Morgan. Each man has one and only one job, but not necessarily according to the order stated above. To help find out who's who, we have a few clues.

1. Clark and Jones are neighbors and drive each other to work.
2. Jones makes more money than Morgan.
3. Clark and Smith are friends and bowl together every Friday.
4. The butcher always walks to work.
5. The only time the policeman ever met he grocer was when he gave him a ticket for speeding.
6. The policeman makes more money than either the druggist or the grocer.

Can you match the man with the job?

Notes and Music

The chairman of the music department of a high school, checking up on the musical activities of his students, came up with the following information: 68 students belong to an orchestra, 72 students belong to a concert band and 25 students are members of a dance band. He also learned that 24 students belonged to both the orchestra and the concert band, nine students belonged to both the orchestra and the dance band, 13 students were members of both the concert and the dance band, and five students belonged to all three musical groups. The chairman's problem: How many students in all were active in musical activities?

High School Hassle

At a certain mythical high school, freshmen always lie, and seniors always tell the truth. One day a teacher met a group of three students -a group that he knew was composed only of freshmen and seniors. The teacher asked the first student whether he was a freshman.

The first student answered the question, but the teacher did not hear the reply. The second student then said that the first student denied being a freshman. Then the third student said that the first student was really a freshman.

From this information, the teacher was able to decide how many in the group were freshmen and how many were seniors. Can you do the same?

Cool Question

Joan asked Fred to check the temperature in their food freezer. As it happens, the freezer had two thermometers, one reading in degrees centigrade and the other in degrees Fahrenheit. Fred glanced at the thermometers and told Joan the temperature.
"Is that centigrade or Fahrenheit?" asked Joan. "Both," answered Fred, "the two readings are the same." What was the temperature in the freezer?

Age Is Relative

When Mr. Adams was asked his age by his son, he replied: "I spent 1/12 of my life in infancy, and then 1/4 of my life in school. Six years after leaving school, I was married and lived another ½ of my life until you were born. I worked hard for 20 years after that and then retired. Since retirement, I have spent 1/4 of my years. That is all I will tell you." How old is Mr. Adams? How old is his son?

Cagey Caliph

In a certain mythical country there lived a Caliph who collected taxes from each of his twelve provinces. The taxes from all the provinces were paid in the same coin of the realm, each coin weighing exactly one pound. On the day the taxes were due, twelve bags of coins were brought to the Caliph's palace, each bag bearing the name of the province, it came from.

Suddenly, a dervish whirled up to the Caliph. "Master," he cried, "one of your princes has cheated you. Every coin from his province is one ounce underweight. His name is..." The dervish said no more as a knife, thrown by an unseen hand, struck him dead.

Now it happened that the only scale in the country accurate enough to measure weights as small as an ounce was a coin-operated one, requiring American pennies.

The Caliph had but a single American penny, so there could be only one weighing.

Nevertheless, the next day a prince's the head rolled -and that of the guilty one.

How did the Caliph find the underweight bag?

Fly Fun

While lying in bed one day. Ken, a mathematics student, noticed a fly in one corner of his room, perched one-third of the way down from the ceiling. At the corner of the room diagonally opposite the fly. Ken noticed a spider, sitting right up against the ceiling. Being a mathematician. Ken quickly figured out how far the spider would have to crawl to reach its prey by the shortest possible route.

Ken's room happens to be nine feet square, and just nine feet from floor to ceiling. Can you solve the problem as quickly as Ken did?

Mind tickler

Cat and Mouse

If five cats can catch five mice in five minutes, how many cats would be needed to catch one mouse in one minute?

Speed Trap

A meticulous professor drives two miles to work every morning. He knows that he must average precisely 30 mph to arrive on time. One morning heavy traffic cut his average to 15 mph for the first mile. The professor quickly calculated the speed he must average for the second mile in order to arrive on time with a 30 mph average for the two miles. How fast did he have to drive?

Match Muddle

You are challenged to position six ordinary kitchen matches so that each one touches every other one directly. If you try to position the matches like the spokes of a wheel, bringing the heads or ends together, you will find that every match does not touch every other match. However, there is a solution. Have you been challenged to a mismatch?

Light Lighter

A sudden thunderstorm overturned the boat in which a girl was rowing across a large lake. The girl managed to climb atop the upturned bottom of the boat, while
- salvaging an old kerosene lamp and some dry matches from the original contents of the boat. As darkness fell, she tried to light the lamp as a signal for help. Unfortunately, there was only a small amount of kerosene in the bottom of the lamp, and the wick was too short of reaching it. With a little scientific ingenuity, she managed to get the lamp lit, however, and summoned help. How did she do it?

Triangle Teaser

How many different triangles can you see in the star-shaped figure below? Be careful: there may be more than you think

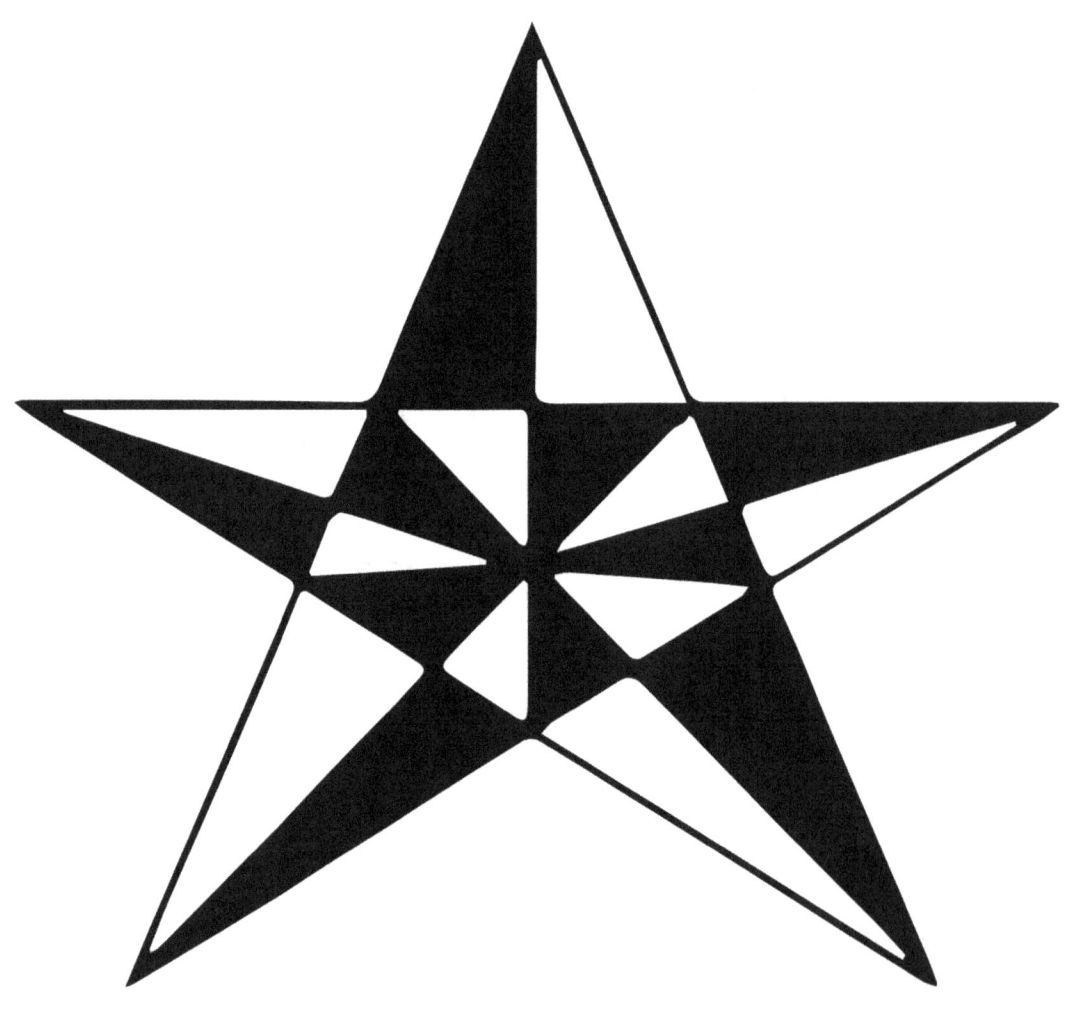

Striking Sounds

A man had a clock that struck the hours and also struck once to mark the half hours.

One night he came home late.

As he opened the door, he heard the clock strike once. Half an hour later, it struck once.

Again, half an hour later, it struck once; and half an hour after that, it struck once more.

What time was it when he came home?

Find the Rule

The numbers from 1 to 10, as listed below, are in a definite order, reading from left to right.

$$8\ 5\ 4\ 9\ 1\ 7\ 6\ 1\ 0\ 3\ 2$$

Can you find the rule which determined the order in which these numbers are arranged?

Crystal Clear

Drawing A below shows six glasses, three of which are filled with water.

By moving only one glass, can you make the six glasses appear as in drawing B?

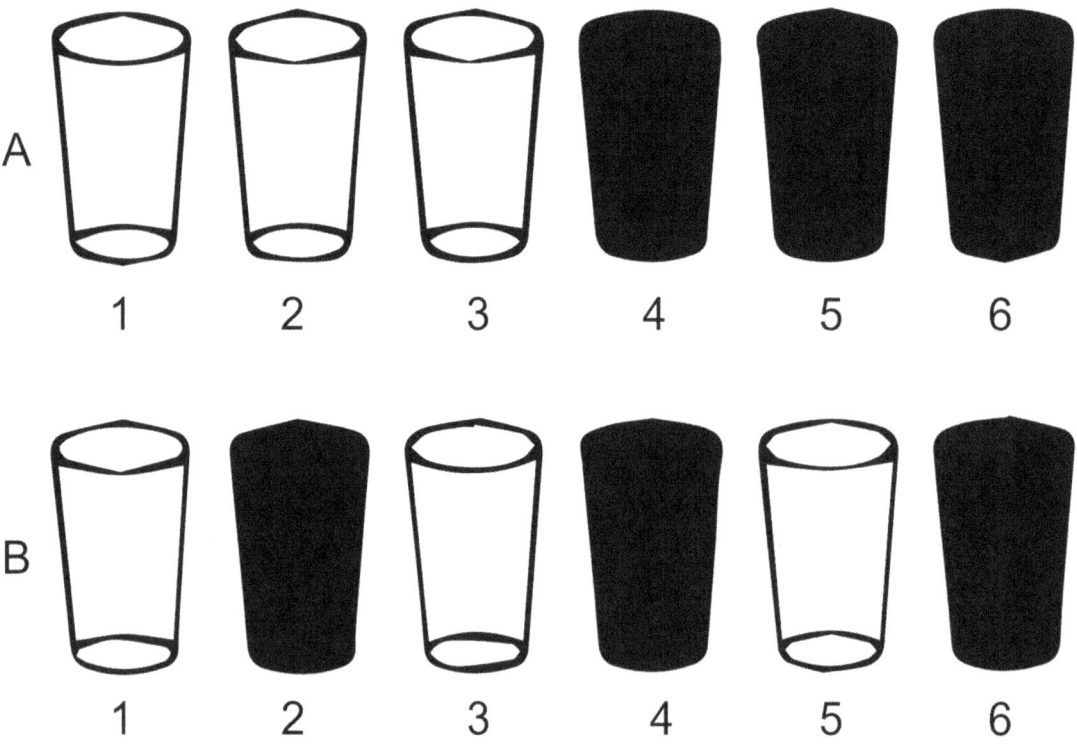

Queen Quiz

Can you place five queens on a chess board so that they govern every square and do not protect each other?

Toaster Teaser

Suppose you have an old-fashioned toaster that takes two slices of bread at once but toasts each slice only on one side. To toast both sides, you have to remove the slices, turn them around and put them back in. If it takes a minute to toast one side, how long will it take you to toast both sides of three slices of bread, assuming that you toast them in the shortest possible time?

Shell Game

You've probably heard of the gambler's old "shell game," but here is a new version. Place three walnut shells in a row, as shown below. Now turn over two shells at a time so that after three moves, all three shells will have their open faces down. If you need more than three moves, you lose

Mind breakers

Incomplete Sentence

Black Bart, the pirate, was sentenced in court on Saturday. The judge announced:
"The prisoner is to be hanged at noon on one of the seven days of next week. He is not to know on which day the hanging will be until the morning of the day of the hanging." The judge added, "If the conditions of this sentence are not met precisely,
the prisoner must go free." Fortunately for Black Bart, the judge was a man known for keeping his word. Black Bart quickly explained to the judge why the sentence
could not be carried out as it had been
started and became a free man.
 Can you determine what Black Bart must have explained to the judge?

Old Teaser

Here is one of the oldest brain teasers there is. It may have originated when a man was actually faced with the problem, solved it out of necessity, and then went on to pose it to his friends.

A man found himself on one side of a river with a fox, a goose, and a sack of grain. He wanted to get to the other side with all of his possessions. A row-boat was available with which to cross the river. However, the man could transport only one possession at a time in the row-boat, and if he left the fox and goose alone together, the fox would kill the goose. If he left the goose and grain alone together, the goose would eat the grain. The man transported everything safely to the other side of the river. His solution is as good today as it was when he used it. What was his solution?

Long Division

In the long division shown below, someone accidentally substituted letters for numbers. If you knew that each letter stands for one of the digits from to 9, could you find all the numbers in the division?

Don't use a trial-and-error method —simple reasoning should give you the answer.

```
              G H I E
          _____
    A B ) C D E B F
          C B
            J E
            A B
            A H B
              A F F
              H B F
              H B F
```

Balancing Act

In the illustrations below, a number of cubes, balls, and cones are balanced against one another on a pair of ordinary balance scales.

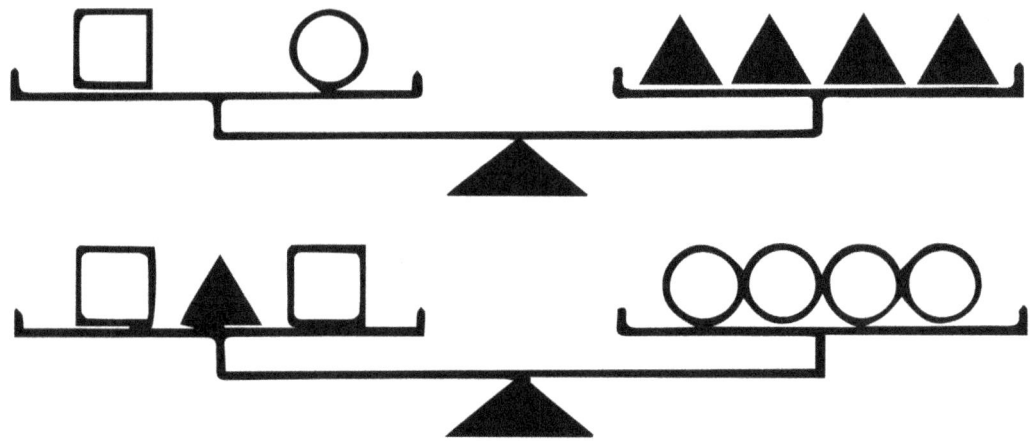

Note that the first balance shows that one cube and one ball weigh as much as four cones, while the second balance shows that two cubes and one cone balance four balls. What is the least number of balls or cones (not cubes) that can balance one cube?

Assume that objects of the same shape are identical. Paper and pencil are not necessary

Blind Multiplication

In the following multiplication, each letter stands for one of the ten digits, but no letter is (zero) or 4. The same letter always represents the same digit.

$$\begin{array}{r} A\,B\,C\,D\,E \\ \times\,4 \\ \hline E\,D\,C\,B\,A \end{array}$$

Can you complete the multiplication?

Round Trip

Mr. Black takes the same train every day, and arrives at his home station exactly at 5 P.M. He is met by his chauffeur exactly at the moment of 5, and is driven home. For the purposes of the problem, assume that no time is lost by Mr. Black in getting into his car, nor is any time lost by the car in making the turn back to Mr. Black's house. One day Mr. Black altered his usual custom. He finished his work early, so he took the train that gets in at 4 o'clock instead of 5 o'clock as usual. When he arrived at the station, he started to walk home but met the chauffeur coming to meet him as usual. He arrived home 20 minutes early. Assuming that the car always travels at the same speed, how long did Mr. Black walk before he met the chauffeur?

Family Party

The following puzzle was supposedly a favorite of Lewis Carroll, the English mathematician, and author of Alice in Wonderland.

The prime minister wants to give a small dinner party. He expects to invite his father's brother-in-law, his brother's father-in-law, his father-in-law's brother, and his brother-in-law's father. If the relationships in the prime minister's family happened to be arranged in the most advantageous manner, what would be the minimum possible number of guests at the party? (Assume that cousin marriages are permitted.)

Direct Dialing

Cute Cora never tells anyone her telephone number. If a would-be boyfriend asks her, she gives him this intriguing reply: "The sum of my telephone number's four digits are 24. The first digit is three times the third, and the second digit is two less than the fourth, and if you're still interested, the fourth digit is one more than twice the third." Cora is vain enough to feel that anyone who is really interested in having a date with her won't be stopped by this test of wits.
Can you give her a ring?

Payment Problem

A well-known citizen walked into a clothing store in his home town to buy a new hat. After choosing one with a $10 price tag, he discovered he didn't have enough money in his pocket to pay for it.

He made the following proposition to the store manager:

"If you will lend me as much money as I have in my pocket now, I will buy the $10 hat."

The manager agreed, and the hat was bought and paid for. Then for the fun of it, the citizen went to another store and repeated his proposition.

This time he bought and paid for a $10 pair of shoes.

At a third store, the same arrangement enabled him to buy a $10 umbrella. When he came out of the store, he had no money left.

How much money was the borrowing citizen carrying when he began his buying spree?

Lewis Carrols Squares

Lewis Carroll is credited with having originated this mind-bender. The problem is to draw the figure below, three overlapping squares, without crossing any lines or lifting the pencil from the paper

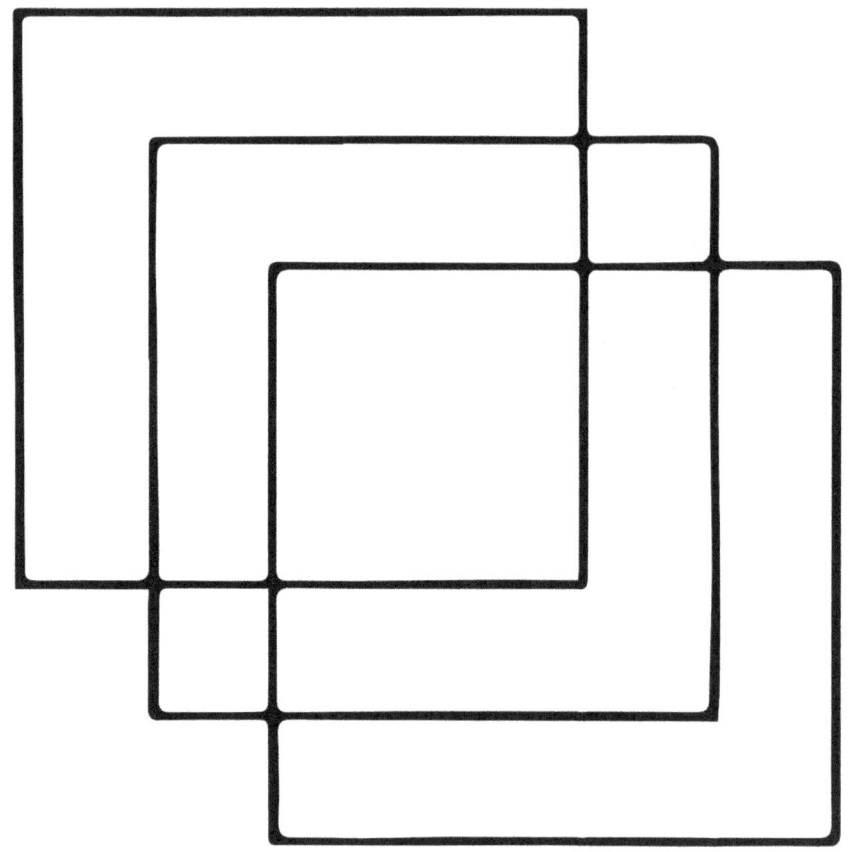

Blind Luck

An eccentric millionaire taught his grandson the following philosophy: "A man must have some luck to succeed. But a good man will make most of his own luck."

In accordance with his philosophy, the millionaire offered his grandson a gift of $100,000 provided the grandson could demonstrate his luck on this test: 500 white and 500 black marbles were to be distributed in two bowls. The grandson was then to select blindfolded one marble at random from either of the bowls. If the marble he selected was black, the grandson would receive the gift. If it was white, he would receive nothing.

The one chance the grandson had to make his own luck was that he could determine the distribution of black and white marbles in the two bowls.

How should the grandson distribute the marbles in order to give himself the best possible chance of winning?

Ailing Alibi

"Shifty Pete" was known to have a grudge against "Sitting Duck Sid." One very cold night in January, Sid was murdered while reading a book in front of the fire in his fireplace. "Shifty Pete" was known to have been in the vicinity at the time and was picked up for questioning. With a manner of complete innocence, he gave the following alibi:

"Sure I was around Sid's house that night. As a matter of fact, as I went by I I went up to the window and cleared the frost off it so I could see in.

"I saw Sid slumped over in front of the fireplace. I got out of there fast, because I knew I would be suspected due to my grudge against Sid.

"That is the truth and the whole truth."

What gave the police good reason not to believe "Shifty Pete"?

Four for Forty-five

Divide the number 45 into four parts so
that when two is added to the first, subtracted from the
second, multiplied by the
third, and the fourth divided by two, the
result of each operation will be the same

Mind Bogglers

Pie Problem

With three cuts, how could you form eight pieces, equal in area, from a pie which is 12 inches in diameter? The individual cuts need not necessarily be straight lines. What would the dimensions of the pieces be?

Pouring Problem

Two men each wished to buy two quarts of wine. One of the men had a five-quart container. The other had a four-quart container. When they reached their local winemaker, they found that he had only two full ten-gallon wine casks. Without measuring instruments and without estimating, the wise old winemaker was able to provide two quarts of wine in each man's container.
He didn't spill a drop in the process. How did he do it?

Plot Problem

A company wishes to build a housing development within the area of a plot in the shape of an equilateral triangle. They wish to have three roads, each perpendicular to one side of the plot, leading to the center of the development. Where should the houses be placed so that the total length of the roads is the least possible, thus conserving roadbuilding expenses?

Generous George

"I divided $16 among my children yesterday," said generous George to an acquaintance. "No child received less than four dollars, and two of my children received the same amount."

"That flimsy information doesn't tell me much," replied the acquaintance. "In that case," continued generous George, "suppose I tell you that the product of the various amounts I gave to each of them equals the number of square inches in a certain number of square feet. That information should tell you how many children I have and how much I gave to each of them."

Generous George's friend was stumped. Can you solve the problem?

Fast Switch

In the next diagram, two railroad cars, A and B, and a switch engine, E, are on the tracks of a railroad yard in the positions shown.

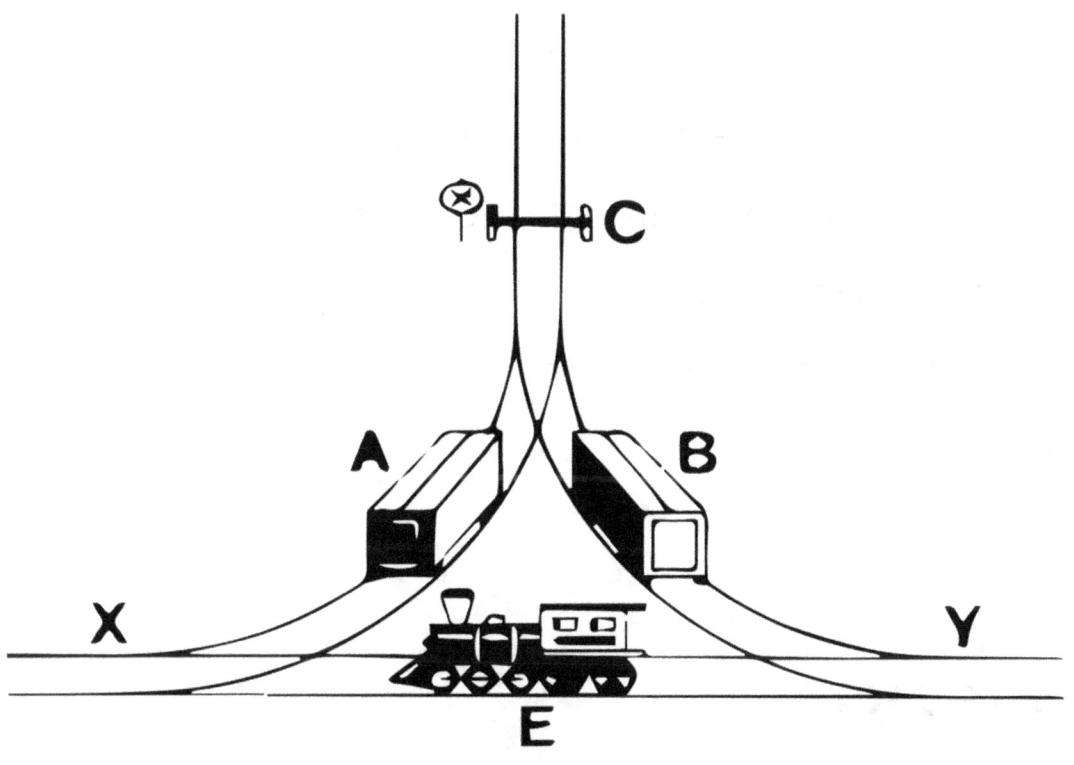

The locomotive engineer running the switch engine is given the following job: to exchange the positions of cars A and B, and to return the switch engine to position E. The engine cannot switch tracks at dead-end C, but one car at a time may do so. It looks simple, doesn't it?

Young and Old

When a mathematician was asked his age this year (1964), he replied: "The square root of the year in which my grandfather was born plus the square root of the year in which my son was born gives the age of my grandfather when he died. If I add 13 to my grandfather's age when he died and divide the sum by 2, I will have my own age at the present time. How old am I?"

Tube Teaser

What is the maximum number of tubular fluorescent lights, every two inches in diameter and 17 inches long, which can be packed into a box having interior dimensions of 18 inches in length, 16 inches in width, and 16 inches in depth? (Hint: Use 1.73 for $\sqrt{3}$.)

Who Dunnit?

Five Suspects were rounded up in connection with a jewel theft which had taken
place in Buckingham Palace on March 15.
Inspector Flynn of Scotland Yard questioned each man and detected the culprit.
Flynn knew one thing which made the
solution possible: Each suspect made one,
and only one false statement. The suspects' statements were:

"Big Sid": I was in Paris the day of the
robbery. "Bugsy" is the guilty man. "Fats"
lies when he says, "Long Stretch" did it. I
am innocent.

"Bugsy": "Big Sid" lies when he says I
did it. I am innocent. "Fats" is not telling
the truth about being in Paris. I never stole
anything in my life.

"Fats": "Long Stretch" did it because he
told me so. I am not the guilty man. I was

in Paris with "Big Sid" on the day of the robbery. "Bugsy" is

"Long Stretch": "Bugsy" lies when he

says "Fats" is not telling the truth. I am

innocent. "Fats" is the man who did it.

"Toad Face" has been in Istanbul.

"Toad Face": I am innocent. "Big Sid" is

innocent too. I have just come from Istanbul. "Fats" was in

Paris with "Big Sid" on

the day of the theft.

Now —Who dunnit?

Candy from a Baby

Two brothers. Max and Sam, buy 20 pieces of candy. They agree on the way to divide the candy between them and to determine who will pay for it.

They will take turns removing pieces from the bag of candy. Each brother must take either one or two pieces on each turn. The one who takes the last of the 20 pieces must pay for all the candy.

Being the elder. Max goes first. With the correct strategy he can make certain that Sam will have to pay for the candy. What should that strategy be? Would it work if there were 22 pieces of candy?

Ant Power

Four black ants can lift as much grain as five red ants. Two red ants and a black ant can lift as much grain as two beetles. Could two beetles and three red ants lift as much grain as one red ant and four black ants? If they could lift more or less, how much more or less?

Tip of the Hat

Five men met at a restaurant for dinner. Three of them brought black hats, and two brought white. Just as the main course was being served, the lights went out. The men hurriedly left the restaurant, each putting on a hat in the dark as he went. None knew the color of the hat he was wearing. Outside, they walked down the street in a single file. Each could see the hats worn by the men in front of him. The fourth in the line said, "I can see three hats in front of me, and I don't know what color I have on."
The second man in line said, "I can see one hat in front of me and I don't know what color hat I have on."
The man at the head of the line said, "I can't see anyone's hat but I know the color of the one, I'm wearing."
What color was it, and how did he know?

Discipline Problem

A father discovered his son floating a silver bowl containing a dozen marbles in the bathtub. Before he could scold him for this, the boy removed the marbles from the bowl dropped them into the water, and asked his father the following question: "Is the water level in the tub higher, lower, or the same now that the marbles are in the water, and the bowl is floating empty on the surface?"

The father was so puzzled by the question that he forgot his annoyance with his son. Can you help him with the answer?

Whose Bananas?

Three castaways and a monkey were marooned on a desert island. During their first day on the island, they gathered a large pile of 50 to 100 bananas. It was agreed that the next morning the bananas would be divided equally among the men.

During the night, one of the three awoke and decided to take his one-third share and hide it so that the others couldn't cheat him. Since there was one banana more than a quantity which could be divided into even thirds, he gave it to the monkey. A short while later a second man awoke and repeated the procedure by taking one-third of the remaining bananas and hiding them. Again there was one extra banana which he too gave to the monkey. Then the third man woke up and decided to take his share. He took one-third of the pile which remained and had one extra which he gave to the monkey.

32 When the men got up in the morning they all noticed that the pile was much smaller, but no one said anything for fear of giving himself away. They divided the remaining bananas three ways and had one extra left over for the monkey.

How many bananas were there in the original pile?

ANSWERS

Timber Land

Here's how the split was done:

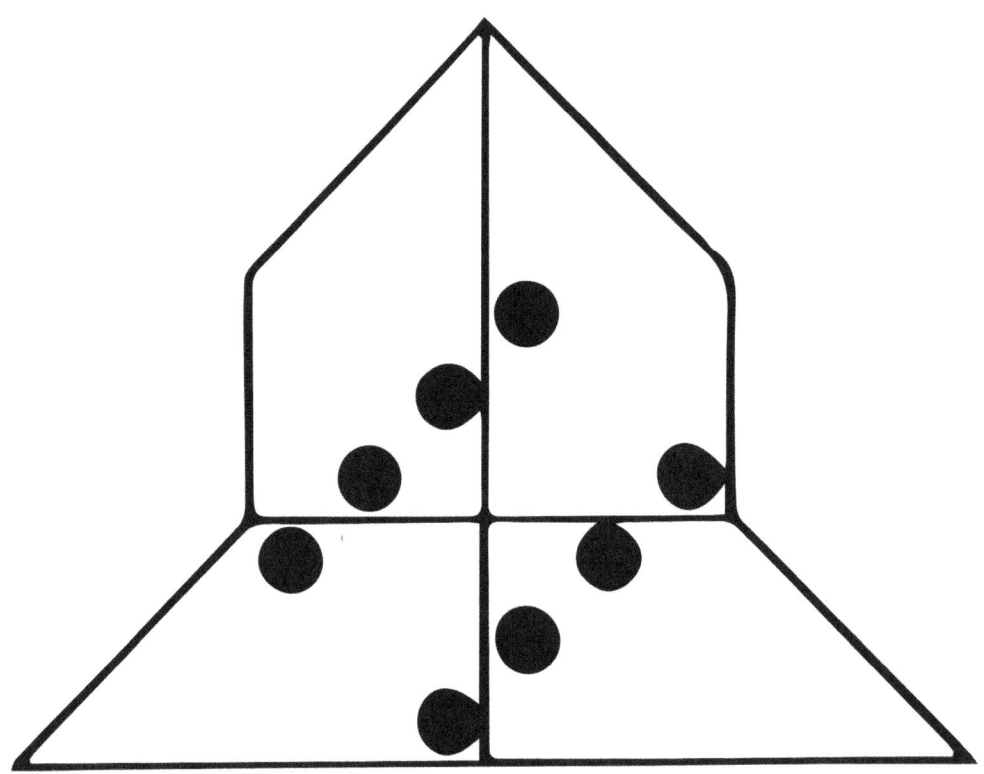

Rare Window

The diagram shows the size and shape of the two windows.

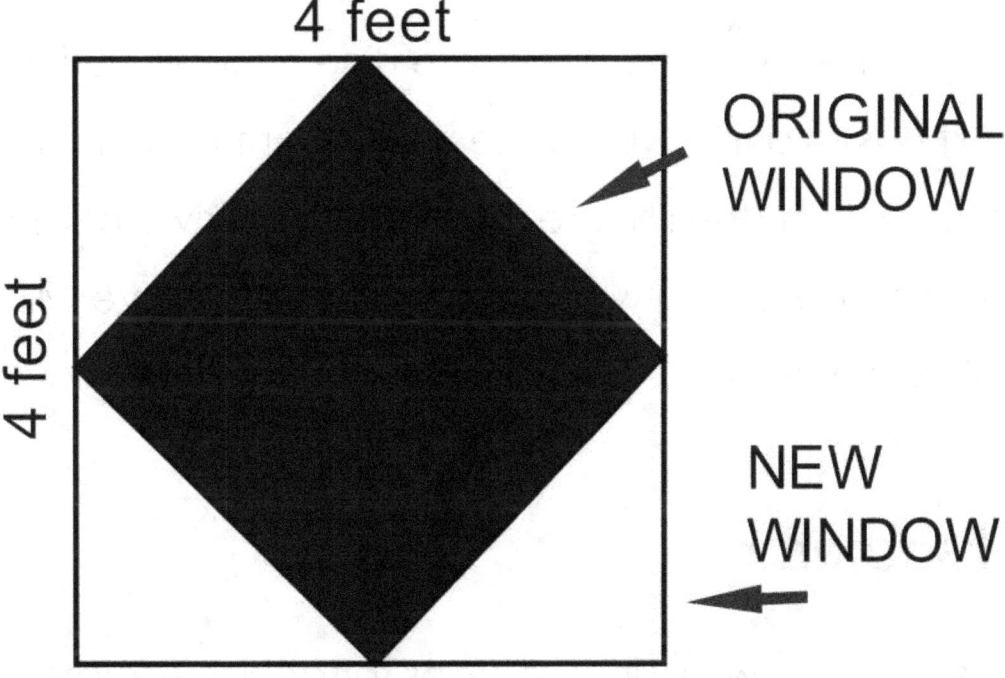

ANSWERS

Bus Stop

Although both buses run at regular ten-minute intervals, they do not come to Melvin's stop at the same time. As stated, Melvin takes the one which comes first once he has reached the bus stop.

What occurs is that the North Centreville bus arrives at the stop one minute after the South Centreville bus. Thus, during every ten-minute period, there are nine minutes during which the next bus to arrive will be the South Centreville bus and only one minute (the minute following the arrival of the South Centreville bus) during which the next bus to arrive will be the North Centreville bus.

The Escalator

In two seconds, I walk up two steps, and the escalator takes me an additional step up, so I advance 3x8 inches or two feet. Therefore, in 20 seconds, I reach the upper floor level, and I have taken 20 steps.

Fractured Fraction

The Fraction is 2/5. If it is reduced to the decimal system, we have (.8 + .2) X .4 = .4. Can you think of any other fractions which meet the requirements of the problem?

Magic Square

To construct a magic square with the numbers from 1 through 9, start by writing the numbers consecutively in three rows as described in the problem.

Then exchange the number in the upper right-hand corner with the number in the lower left-hand corner, and switch the number in the upper left-hand corner with the number in the lower right-hand corner, as follows:

```
9 2 7
4 5 6
3 8 1
```

Next, move all the numbers one space counterclockwise around the square, leaving the center number in position:

$$\begin{array}{ccc} 2 & 7 & 6 \\ 9 & 5 & 1 \\ 4 & 3 & 8 \end{array}$$

The resulting rows and columns add up to 15 in any direction. This procedure works for any series of nine consecutive integers, although the sum will be different for each series.

Cannonball Express

The trains are approaching each other at a relative speed of 60 mph. Therefore they will pass each other after 10 minutes. Since the fly is flying steadily at a speed of 60 mph, it will cover exactly ten miles during the ten minutes it takes the trains to meet.

Tick-Tock Teaser

The watch that does not run at all tells the right time every 12 hours, or twice a day. On the other hand, the watch that loses one minute a day will not show the correct time until 720 days later, when it will be exactly 12 hours behind schedule. Similarly, the watch that gains a minute a day will also show the correct time 720 days later, when it will be 12 hours ahead of schedule.

Thus, the watch that does not run at all is most likely to show the correct time; the watches that gain or lose a minute a day are equally likely to be wrong.

Pioneer Puzzle

The two children first stepped into the boat and rowed across the river. One got out and waited on the bank while the other child rowed back. The child got out; his mother got in the boat, and she rowed across by herself. She then got out on the other side, and let the child who had been waiting there step into the boat. That child rowed back, picked up the other child, and the two children rowed back once more to the far bank where their mother was now standing. They then repeated this procedure to bring their father across the river, and the entire family continued on its way.

Busy Worm

It takes the worm 25 days to reach the top of the telephone pole.

After each day and night on the pole, the worm has gained only one foot.

After 24 days and nights, the worm is 24 feet up the pole.

The following day - the 25th day- the worm climbs six feet, reaching the top of the pole.

Lightning Multiplier

The young mathematician immediately saw that all the numbers from 1 to 100 could be arranged in pairs, as follows: 1 and 100, 2 and 99, 3 and 98, etc., until 50 and 51. Since the sum of each pair is 101, and there are 50 such pairs, the problem becomes one in simple multiplication:

50 X 101 = 5050. If you want to check this answer, you can always take the trouble of adding all the numbers from 1 to 100.

ANSWERS

Collective Farm

If the plot of land is divided as in the diagram below, each son will receive the same amount of land in the same shape.

Happy Hunter

The hunter's statement was, "I shall die by fire." Then he could not be killed by fire since the statement would be true, nor could he die by poison since the statement then would be a false one.

Does 1 = 2?

The fallacy occurs in step (4), where both sides of the equation are divided by (a - b).

Since a = b, (a - b) has a value of zero. Dividing any number by zero gives an indeterminate result, and is not permitted under the rules of ordinary algebraic manipulation.

ANSWERS

Number Stumper

The problem cannot be solved unless the numbers 1 and 10 are placed in the two center circles.

The reason for this is hidden in the fact that 1 and 10 are consecutive to one number only, while all other numbers 2 through 9 are consecutive to two numbers apiece. One solution is given below.

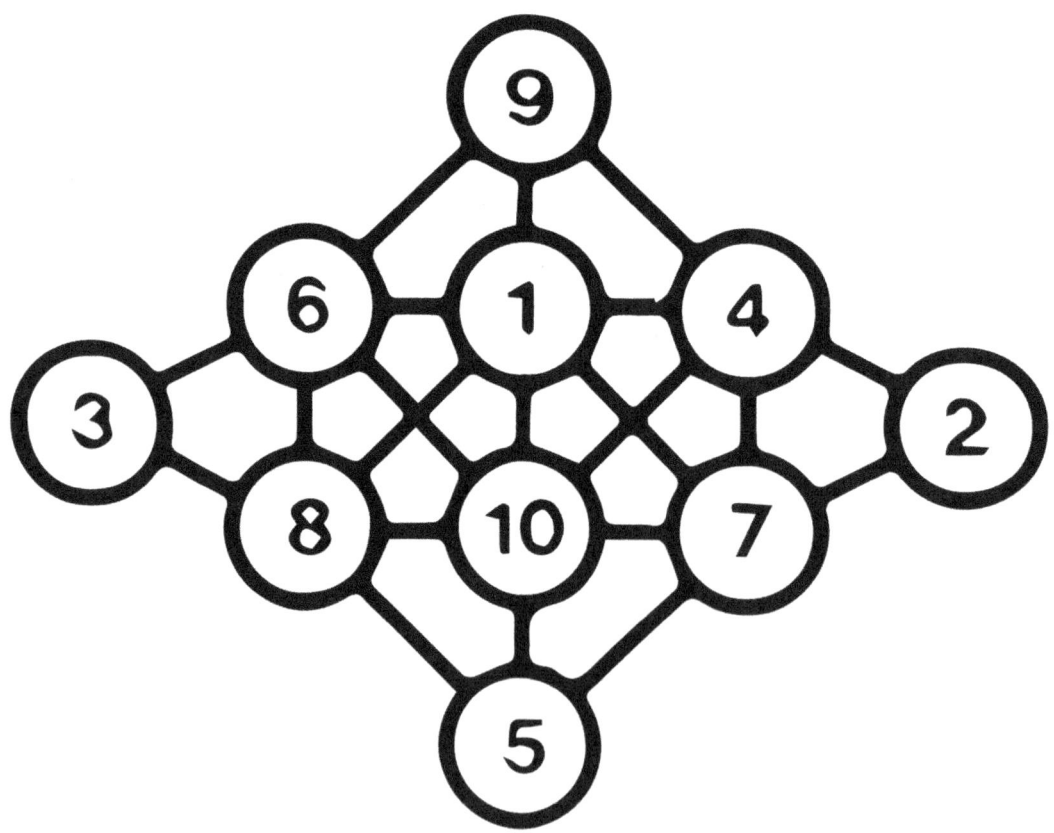

Wet Solution

The chemist dissolved a total of 10 gallons of harmless red dye into various parts of the lake. Two weeks later, after the dye had had time to circulate, he took water samples from several parts of the lake. From these, he determined the average content of dye in each gallon of lake water. From the proportion of the original 10 gallons of dye to the dye content measured in the lake water, he was able to determine how many gallons of water the lake must have contained.

Weighty Problem

First, put three marbles on each pan of the balance scale. If the two loads balance, then you weigh the remaining two marbles. The heavier of these two will tip the scale, revealing its presence in two weighings.

If the first load of six-marbles is unbalanced, pull one marble off each pan. If the remaining loads do balance, weigh the two marbles pulled off the pans. This weighing will thus find the heavier marble.

If the remaining loads do not balance, weigh the two marbles on the heavier side to find the heavier marble. Thus, the heavier marble can be found in no more than three weighings.

Who's Who

We know that neither Clark nor Jones can be the butcher because the butcher walks to work (clue 4), and we know that Clark and Jones drive (clue 1).

One of them has to be the grocer because the grocer has a car (clue 5). If one of the drivers is the grocer, then the other can only be the druggist since the grocer met the policeman only once in his lifetime (clue 5).

We now know that Clark and Jones hold the occupations of druggist and grocer, but we do not know who holds which. The fact that Jones makes more money than Morgan (clue 2) tells us that Morgan is the butcher, for Jones is either the druggist or grocer, and Morgan cannot be the policeman (clue 6). Since Morgan is the butcher, then the policeman must be Smith, by elimination.

It then follows that Clark is the druggist, since Clark is a friend of Smith, the policeman (clue 3), and Clark cannot be the grocer since the grocer only saw the policeman once. This leaves Jones as the grocer.

Notes and Music

The problem can be easily visualized if we think of the three musical groups' membership as split up among three overlapping circles, as in the diagram below. The circles are marked O for orchestra, C for concert band, and D for dance band. We first place "5" in the center portion common to all three circles since five students belong to all three groups. Then, since there are 13 in the concert and dance bands, there are 13 - 5 = 8 students in both bands who do not belong to the orchestra.

Thus we mark "S" in the appropriate section. Similarly, there are 9 - 5 = 4 students in the orchestra and dance band who do not also belong to the concert band; and there are 24 - 5 = 19 students in the orchestra and concert band who do not also belong to the dance band. We now have 4 + 5 + 19 = 28 members of the orchestra who are also active in other musical groups, so that 68 - 28 = 40 students play only in the orchestra.

Similarly, we find that 72 - 32 = 40 students play only in the concert band, and 25 — 17 = 8 students play only in the dance band. We now find that every student is accounted for once and only once in our little diagram. Therefore, 67by simply adding the number of students in each section of the diagram we get 40 + 40 + 8 + 19 + 5 + 4 + 8 = 124 students musically active in the high school.

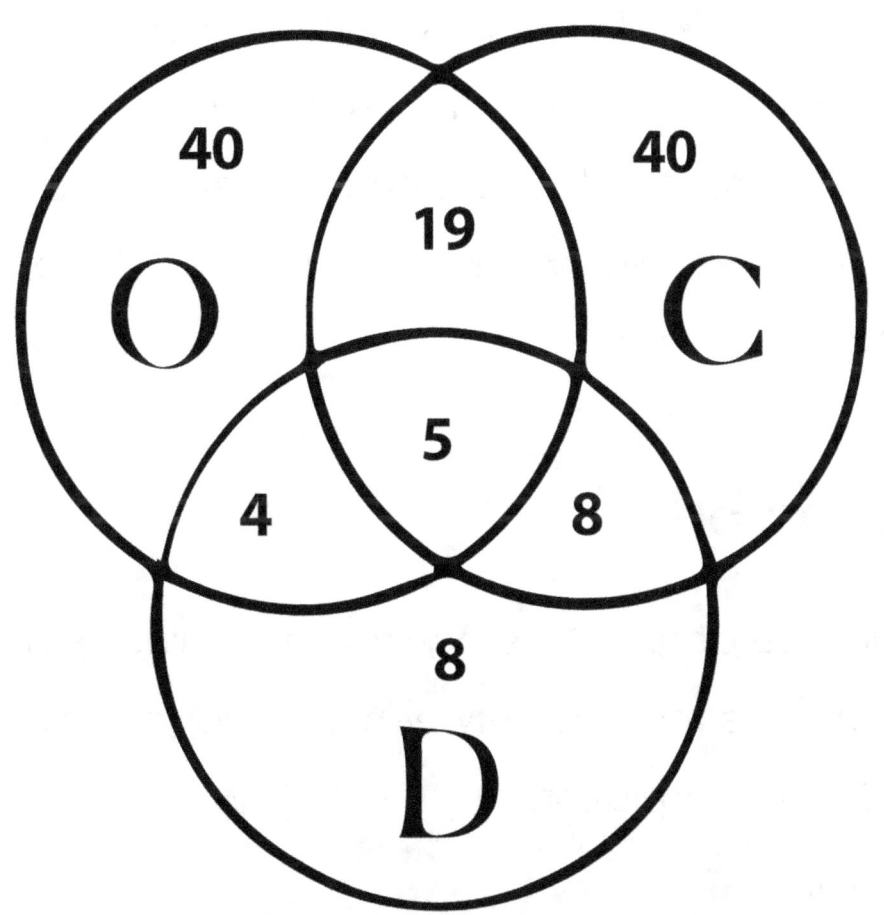

High School Hassle

The first student would have to answer "no" to the teacher's question, whether or not he was, in fact, a freshman. Therefore, the second student's answer was a truthful one. This means that the second student is a senior. Now, if the third student's statement is true, then the third student is a senior, and the first student is, in fact, a freshman. On the other hand, if the third student is lying, then the third student is a freshman, and the first student is a senior.

Thus the first student or the third student could be a freshman, but not both. Accordingly, the group of three students must consist of one freshman and two seniors.

Cool Question

Five centigrade degrees are equivalent to nine Fahrenheit degrees, and the relation between the two scales is expressed by the equation

C = 5/9(F-32).

For the two readings to be the same, C and F must both have the same numerical value, say X.

Substituting in the equation, we get:

X = 5/9(x-32)

so that

X=-40

Thus the temperature in the freezer was -40°F. and C: the only temperature which has the same numerical value on both scales.

Age Is Relative

Mr. Adams had accounted for every year of his life.

The total consisted of: 1/12 + 1/4 + 1/18 + 1/4 parts of his life, and 6 + 20 years. This amounts to 23/36 of his life, plus 26 years. Thus his whole life to date 36/36, minus the 23/36, is 13/36. This 13/36, we know to be 26 years.

The full 36/36 of his life, then, must be 72 years.

From the time his son was born, Mr. Adams lived 20 years, plus 1/4 of his total years, 18 years.

This amounts to 38 years, his son's age.

ANSWERS

Cagey Caliph

The Caliph placed one coin from the first province on the scale, two from the second province, three coins from the third, and so on, for all twelve provinces. Then he measured the combined weight of these 78 coins.

The number of ounces less than the expected 78 pounds would then be the same as the number of the province, pointing the finger at the guilty prince.

Fly Fun

The solution is quickly found by mentally "unfolding" the walls and ceiling of the room and laying them out flat, as in the diagram above. A straight line drawn between the spider and fly on this diagram represents the shortest distance between the insects; any other path would be longer.

Note that lines A and B are actually the same corners of the room. Accordingly, the distance from the spider to fly is the hypotenuse of a right triangle whose legs are 9 and 12. By the Pythagorean theorem, this length is the square root of 9^2 of 12^2, or 15. Hence the spider must crawl 15 feet to reach its dinner if it were able to solve the problem.

ANSWERS

Cat and Mouse

Five cats would be needed. If five cats catch five mice in five minutes, five cats catch mice at a rate of one a minute.

Speed Trap

The professor could not possibly make it on time. A trip of two miles at an average speed of 30 mph will take four minutes. The professor averaged 15 mph for the first mile, which means the one mile took four minutes to drive. Thus he could not possibly make two miles in four minutes.

Match Muddle

If the matches are positioned as shown, every match touches all five other matches.

Light Lighter

The girl dipped the lamp into the water until the kerosene container at the bottom of the lcmip was full of a combination of kerosene and water. She was then able to light the lamp since the kerosene - lighter in weight than the water- rose to the top and reached the wick.

ANSWERS

Triangle Teaser

There are 85 different triangles in the figure. You can count them by first counting the triangles which do not contain any other triangles. You'll find 20 of these. Then count all the triangles that contain two triangles, of which there are also 20. Next, start at each point of the star and count the triangles that contain three triangles (10), four triangles (10), six triangles (10), eight triangles (10), and 14 triangles (5). This gives a total of 85 triangles and accounts for all the triangles in the star.

Striking Sounds

He heard the last stroke of midnight as he opened the door. The clock then struck once at 12:30, 1:00, and 1:30 A.M.

Find the Rule

The answer is obvious once you see it! The order of the numbers is determined by spelling out each number in letters - eight, five, four, nine, one, seven, six, ten, three, two- and arranging them in alphabetical order: Simple, isn't it?

Crystal Clear

Simply empty glass number 5 into glass number 2.
One possible solution is pictured above.
Can you think of others?

Shell Game

First, turn over shells 2 and 3. Then turn over 1 and 3, and finally 2 and 3 again. All the shells will then have their open face down. The same result can be achieved by turning over 1 and 2, 1 and 3, and 1 and 2 in sequence.

ANSWERS

Toaster Teaser

You can have your toast in only three minutes. The first minute you toast one side of slices A and B. The second minute, you toast the other side of B and one side of C.
The third minute you toast the second sides of slices A and C.

Incomplete Sentence

Black Bart presented the following reasoning to the judge: "They obviously cannot hang me next Saturday, the last day of the week, because if I were still alive Friday afternoon, I would know that Saturday had to be the day since it was the only day left. This would violate the sentence since I would know what the hanging day was to be before it actually arrived.

 Therefore Saturday is ruled out. This brings me to Friday. But if I were still alive on Thursday afternoon, I would know that Friday has to be the day.

Thus, they can't wait until Friday, and it is ruled out. Thus, Thursday becomes the last possible day for the hanging. But if I were still alive on Wednesday afternoon, I would know that the day must be Thursday." Black Bart continued this way until Sunday, the next day, was shown to be the only possible day. He pointed out that it, too, could not be the day because he knew now, the day before, that the hanging must be the next day, and since he knew ahead of time, they had to release him.

Old Teaser

The man first rowed the goose to the other side of the river, leaving the fox and grain behind. He returned to the original point and transported the fox to the other side, where he left it while bringing the goose back to the first side with him. He left the goose on the first side and carried the grain to the other side, where he left it with the fox. He then returned alone to the first side, where he picked up the goose and brought it to the other side where the fox and grain were waiting.

Long Division

We know that H is 1, because in the second multiplication, H x AB = AB.

F is 0 because in the third subtraction HB-- FF = HB.

B is 5 because in the third multiplication the product of Ix B ends in 0, and thus has 5 as a factor.

The product of E x B, in the fourth multiplication, also has 5 as a factor, since it too ends in 0.

This means that either both E and I must be 5, or B, the common factor, is 5. Since E and I are different numbers, B is 5.

Next, we reason that A is 2 because I x AB = AFF, and F is 0. This can be rewritten as I x A5 = A00, and we find that A = 2 is the only figure that will work.

Thus the divisor AB is 25.

I is 8 because in the third multiplication I x 25 = 200.

G is 3 because in the first multiplication, G X 25 is a two-digit number (CB) ending in 5, which means that G must be an odd number less than 4 and greater than 1 (H is 1). C is 7 because in the first multiplication 3 X 25 = CB, or 75.

J is 4 because in the second subtraction JE - AB = AH and A is 2 (also, E is larger than B since E − B = 1). E is 6 since in the second subtraction E - 5 = 1. Finally, D is 9 because in the first subtraction D - 5 = 4.

This completes the division, as shown below.

```
          3186
        _____
    25)79650
       75
       ___
         46
         25
         ___
          215
          200
          ___
           150
           150
```

Balancing Act

Think of the two weighings added together so that the sum of the objects on the left-hand pans, three cubes, a ball, and a cone, balance against the sum of the objects on the righthand pans, four cones, and four balls.

This arrangement is shown in the drawing below.

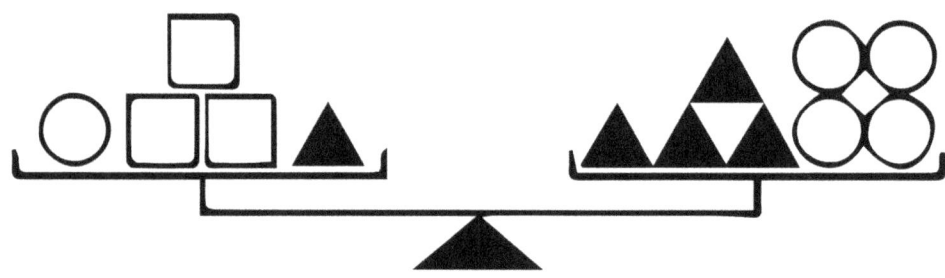

If you then take away a cone and a ball from each pan, you will have three cubes in the left pan and three cones plus three balls in the right pan. This means that one cube will balance one cone plus one ball.

Blind Multiplication

A can't be 1 because 4 X E = A so that E would have to be a fraction. A can't be 3 or greater because the last multiplication would then give a six-place answer. Hence, A must be 2.

Therefore the first multiplication faces us with the question: 4 times what number ends in 2?

Thus E must be 3 or 8. But E can't be 3 since, in the last multiplication, 4 X A = E. We know that A is 2, so E must be either 8 or 9. Therefore E is 8. We now have:

$$\begin{array}{r} 2....8 \\ \times 4 \\ \hline 8.....2 \end{array}$$

We know that the product of 4 X B cannot have a carryover, because it does not show up in the last multiplication. Thus B can't be greater than 2, and since A is 2, B must be 1.

Now, 4 X D must yield a product ending in 1. Taking into account the carryover of 3 from product 32 in the first multiplication, we see that D must be 7. The rest is easy: Since the product of 4 X C yields a carryover of 3, C must be 8 or 9, but we know E is 8, so C is 9.

This completes the multiplication.

Round Trip

The chauffeur, meeting Mr. Black at some point between the station and the residence, saves 20 minutes from his usual trip by not being obliged to proceed from the point where Black is met to the station and then make a return trip to that point. In other words, the chauffeur saves the run of double the distance from the point where Black is met, to the station. The saving, amounting to 20 minutes in all, means a saving of two 10 minute runs.

Therefore, the chauffeur met Black 10 minutes before he would usually arrive at the station. Since he usually arrived at 5 P.M., he met Black at 4:50 P.M. Since Black arrived at the station at 4 P.M. and was met at 4:50 P.M., he had walked 50 minutes.

Family Party

One guest would fulfill all the requirements if the host's family is related as follows: The host's mother has two brothers, brother 1 and brother 2. The host has a brother who married the daughter of brother 1, a cousin. The host also has a sister who married the son of brother 1. The host himself is married to the daughter of brother 2.

Accordingly, brother 1 is the host's father's brother-in-law, the host's brother's father-in-law, the host's father-in-law's brother, as well as the host's brother-in-law's father. This would make brother 1 the sole guest at the party. A diagram may help to visualize the relationships.

Direct Dialing

The sum of the four digits is 24, which can be indicated as $A + B + C + D = 24$, where A, B, C, and D are the first, second, third, and fourth digits of her phone number, respectively.

From Cora's statements we know that $A = 3C$ $B = D - 2$, and $D = 2C + 1$.

Substituting for D, we get $B = (2C + 1) - 2$, or $B = 2C - 1$.

Now substituting C for A, B, and D in the first equation, we get $3C + (2C - 1) + C + (2C + 1) = 24$.

Thus $8C = 24$ and $C = 3$.

Substituting this value into the other equations, we get $A = 9$, $B = 5$, and $D = 7$.

Dial 9537 and hope the line won't be busy.

Payment Problem

The citizen had $8.75 when he began his buying.

This can be determined by working back from the third purchase: He must have had $5 and borrowed $5 to buy the $10 umbrella. To have had $5 left after making his second $10 purchase (shoes), he must have had $15, including his second loan to buy the shoes.

The loan was half of the $15, so he must have entered the second store with $7.50. To have $7.50 left after his first acquisition (hat), he must have had $17.50, including the first loan.

Half of this is his original money — $8.75.

Lewis Carroll's Squares

The squares should be drawn, as shown below. In actuality, the lines should touch, but not cross, where the intersections are indicated.

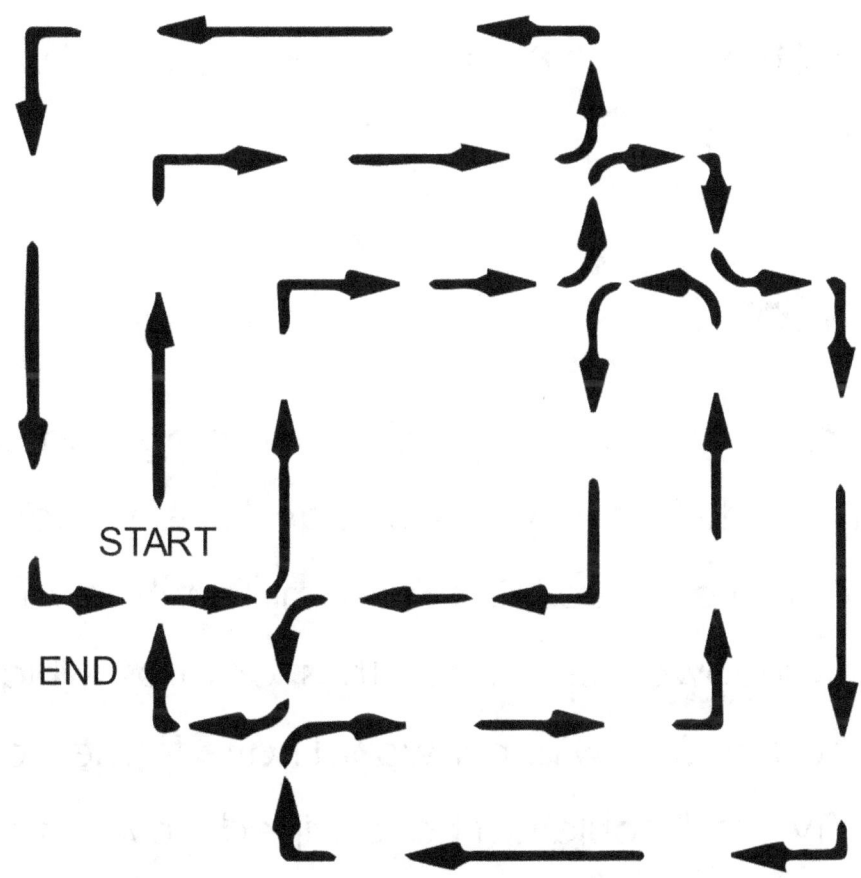

Blind Luck

The grandson should place one black marble in one bowl and all the rest of the marbles in the other bowl. He is sure to win if he picks his marble from the first bowl.

If he picks from the second, he will have a 499 out of 999 chance.

In all, he will have a 1,498 out of 1,998 chance of winning, or almost 3 out of 4.

Ailing Alibi

When there is warm air on one side of a glass and cool air on the other, condensation will occur, and frost will form on the warm side. You can see this, for example, when you breathe on a cold windowpane indoors. Thus, any frost which might have formed on Sid's window would have formed on the inside."Shifty Pete" could not have wiped it away from the outside as he said he did.

Four for Forty-five

Call the four parts of 45 A, B, C, and D:

$A + 2 = B - 2 = C \times 2 = D/2$.

Evaluate each part in terms of

A. $B = A + 4$; $C = (A + 2)/2$; $D = 2(A + 2)$ or $2A + 4$.

Use these expressions to represent all four parts of 45:

$(A) + (A+4) + (A + 2)/2 + (2A + 4) + 45$;

$4A + A/2 + 9 = 45$;

$9A/2 = 36; 9A = 72$.

$A = 8$. $B = 12$, $C = 5$, $D = 20$.

The identical result of the four operations is 10.

Pie Problem

The three cuts required to make eight pieces of equal size are shown below.

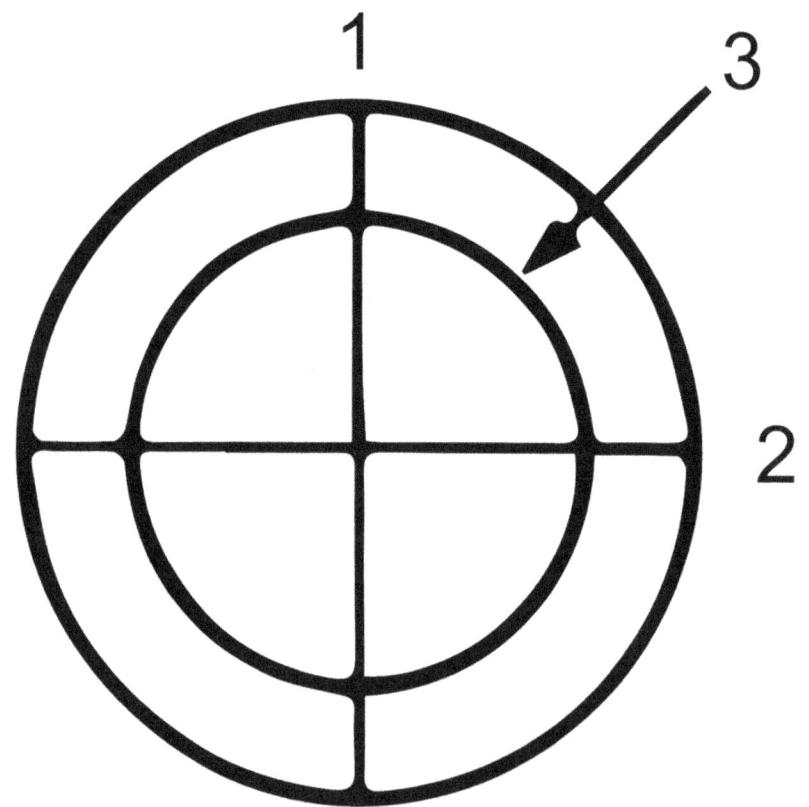

For the eight pieces to be of equal area, the inner circle enclosed by cut No. Three must be equal in area to the remainder of the pie.

The area of the whole pie is 36π square inches since the radius is six inches and the area = πr^2.

Thus, the area of half the pie — the inner circle — is 18π square inches.

By substituting back into Area = πr^2, we have $18\pi = \pi r^2$ for the inner circle.

Then $18 = r^2$, and $\sqrt{18} = r$.

Thus, to cut eight pieces of equal area, cuts No. 1 and No. 2 must divide the pie in fourths and cut No. 3 must be a circle with a radius equal to $\sqrt{18}$.

Pouring Problem

Call the two ten-gallon (40 quarts) containers A and B, The desired result of two quarts in the four-quart container and two quarts in the five-quart container without spilling is reached with the nine pourings shown below.

	A (40qts.)	B (40qts.)	4 qts.	5 qts.
Fill 5 qt. can from A	35	40	0	5
Fill 4 qt. can from 5 qt. can	35	40	4	1
Empty 4 qt. can into A	39	40	0	1
Empty 5 qt. can into 4 qt. can	39	40	1	0

Fill 5 qt. can from A	34	40	1	5
Fill 4 qt. can from 5 qt. can	34	40	4	2
Empty 4 qt. can into A	38	40	0	2
Fill 4 qt. can From B	38	36	4	2
Fill A from 4qt.can	40	36	2	2

Plot Problem

Let the triangle below be the plot of land. Let each side have a length b and place the construction site P anywhere within the triangle. Designate the length of the three perpendiculars to P as X, y, and z. Then the area of triangle APB = bz/2; the area of triangle APC = bx/2; the area of triangle BPC = by/2. Therefore, bx/2 + by/2 + bz/2 = bH/2 where H equals the altitude of ABC.

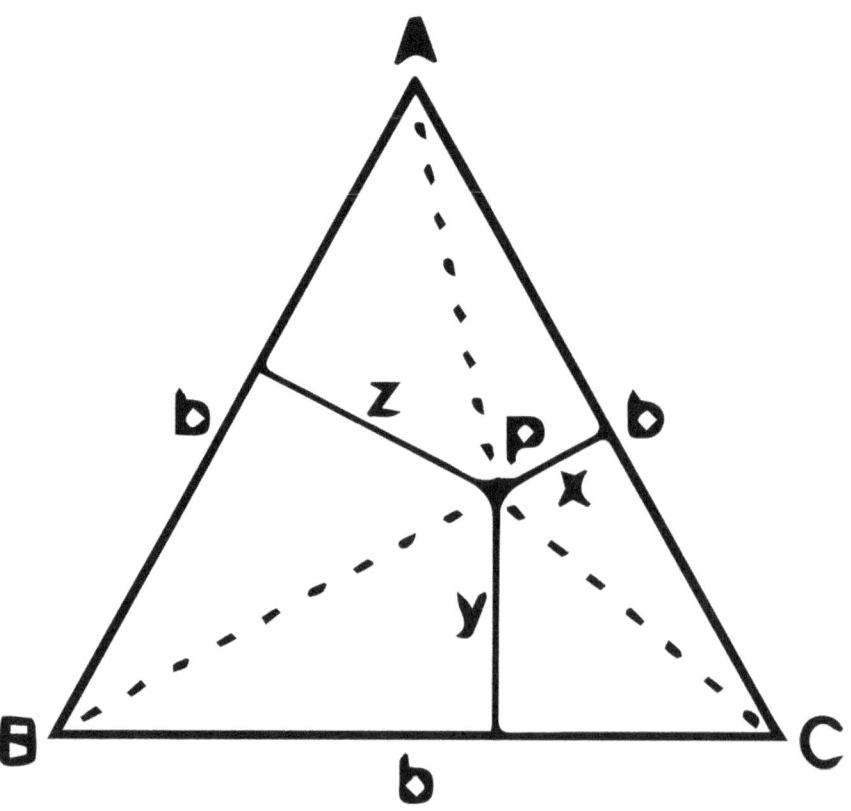

Now multiply by 2 and factor out b, obtaining x + y + z = H. Since H is constant, the sum of the perpendiculars from any point within the equilateral triangle is constant. Thus the development can be built anywhere in the triangle; the amount of road construction will be the same.

Generous George

George used the word "among," which implies that he has more than two children; otherwise, he would have used the word "between." Since no child received less than four dollars and the total parceled out was $16, George could have had either three or four children, but not more than four. But since "two of them received the same amount," the other child (or children) must have received more than four dollars. This means that George could not have four children since the sum of their gifts would then add up to more than $16. Hence George must have three children. We now know that the three gifts must have been either four dollars, four dollars, and eight dollars, or four dollars, six dollars, and six dollars. No other combinations fill the requirements.

ANSWERS

If we call the three gifts a, b, and c, then their product must be some multiple of 144 (number of square inches in 1 square foot), or a x b x c = 144 x n, where n = 1,2,3, ... etc.

But if n is larger than 1, 144 X n will be larger than the product a X b X c in either of our two possible solutions. Hence a X b X c = 144, which only holds true for the second solution, so that the three gifts were four, six, and six dollars.

Fast Switch

Since the engine can move forward or backward, and the problem is symmetrical, it doesn't matter which way the engine starts off.

Suppose the engineer first proceeds to car A.

There, he hooks up to car A and pulls it to point X, where he reverses direction and pulls it to point Y, where he unhooks. He then proceeds to car B and pushes it onto the switch at C. He next takes a long way back to point C, approaching B from the other track. He now hooks on to car B, pulls it past A's original position, down to point X, where he reverses and pulls B to hook up to A at point Y.

The engineer then pulls both of the cars back to X, reverses, and pushes the cars back up to C, where car A is unhooked. He then pulls car B to A's original position, unhooks, and proceeds around and up to switch C, where he hooks onto A and pulls it onto B's original position.

The job completed, he returns the engine to point E and leaves for a long vacation.

Young and Old

The only year with a perfect square root that fits his grandfather's birth date is 1849, of 80 1which the square root is 43. (The next smaller perfect square is 1764, an impossible birth date for his grandfather.) The next year that is a perfect square is 1936 (square root is 44), which must be the year in which his son was born, because the perfect square after that, 2025, is an impossible birth date. The sum of the two possible square roots is 87, the age of his grandfather when he died. Adding 13 to this gives 100, which divided by 2 gives 50, the mathematician's present age.

Tube Teaser

Eight layers of 8 tubes could be placed in the box, each tube directly above the one below. But if 8 are placed in the bottom layer, 7 in the second, 8 in the third, etc., some of the wasted space between bulbs are utilized. This way, the box holds nine layers, five of 8 tubes, and four of 7 tubes, total 68.

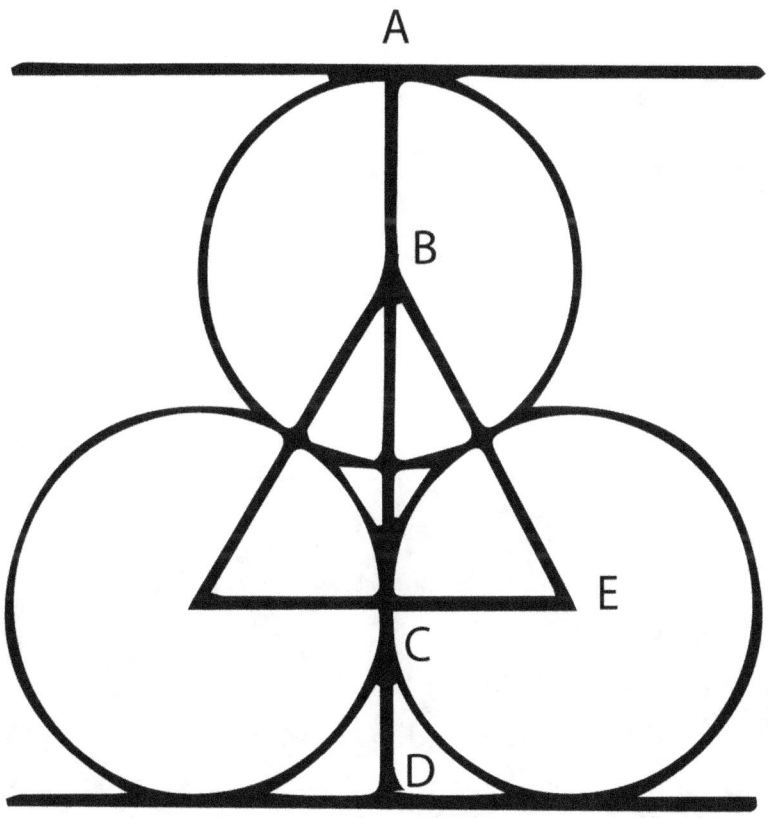

ANSWERS

That nine such layers really fit can be proved, as follows:

The drawing shows the triangle formed by the centers of 3 touching tubes.

AB = 1, CD = 1, and BE = 2. BC = $\sqrt{3}$ (by Pythagorean Theorem, $(BC)^2 = (BE)^2 - (CE)^2$).

Using $\sqrt{3}$ = 1.73, AD = 1 + 1.73 + 1 = 3.73 inches.

Each layer is 1.73 inches higher than the one below.

All nine layers are 2 + 8 (1.73) = 15.84 inches within the 16-inch height.

Who Dunnit?

"Bugsy" said, "I am innocent/' and "I never stole anything in my life." The first cannot be false without the second being false. Since he made only one false statement, the first must be true — that he is innocent.

Therefore, "Big Sid" makes his false statement when he says that "Bugsy" did it. Accordingly, the rest of "Big Sid's" statements must be true. He says that he is innocent and that "Fats" lied in saying that "Long Stretch" did it.

Thus, "Long Stretch" must be innocent, and the rest of the statements made by "Fats" must be true, including his statement that he himself is not guilty. Therefore, the false statement made by "Long Stretch" is the one that accuses "Fats." Also (according to a true statement by "Fats"), "Bugsy's" false statement was the one claiming that "Fats" was lying about being in Paris.

ANSWERS

This leaves only "Toad Face." We know the other four are innocent, and we have identified the one false statement each made. The last three, of "Toad Face's" statements are confirmed as true by true statements made by the others. Thus, his one false statement must be that he is innocent. Therefore — "Toad Face" dunnit.

Candy from a Baby

To make Sam take the last piece of candy. Max must take the next-to-last one -the 19th. He can do this if his first selection leaves one more than an integral multiple of three pieces. Then, by combining with Sam's selections so that together they take three pieces on each pair of turns. Max can be sure to get the next-82to-last. If Sam took two. Max would take one; total, three. If Sam took one. Max would take two. Thus, by taking one piece to start,
leaving(6) 3 -f 1 = 19.

Max will be assured of taking the next-to-last piece, and Sam must take the 20th. Since Sam will have to pay, he might as well, take two pieces each time, obtaining 13 pieces while Max gets seven.

Max could not apply this strategy with his initial choice if there were 22 pieces since this represents one more than an integral multiple of three, to begin with. Therefore Sam, going second, he can combine his selections with Max's to proceed in multiples of three and take the 21st piece, forcing Max to take the 22nd.

Ant Power

We know that four black ants can lift as much as five red ones. Two beetles can lift as much as a black ant and two red ants. In the question posed, substitute the equivalent strength of black and red ants for the two beetles. This would be one black ant and two red ants. The question then becomes: Can one black ant and five red ants lift as much as four black ants and one red ant? We know that the five red ants on one side are equal to the four black ants on the

This leaves one black ant on one side and a red ant on the other. We know that a black ant can lift 25 percent more than a red ant because four black ants can lift as much as five red ants. Thus the group of two beetles and three red ants would lift more than the group of one red ant and four black ants. The margin of difference would be 25 percent of the total amount that one red ant could carry.

Tip of the Hat

The fourth man could not see three black hats or two white hats among the three in front of him, or he would have been able to identify the color of his own. Thus, he must see two black hats and one white hat. The second man knows this and also knows that if he sees the one white hat in front of him, a black hat must be on his own head.

The first man realizes that since the second man could not determine the color of his own hat, he — the first man — must be wearing a black hat. (If the first man had had a white hat, the second man would have known the color of his own hat.)

Discipline Problem

The water level will drop. In the bowl, the marbles displace an amount of water equivalent to their weight. When they are actually in the water, they displace an amount of water equivalent to their volume. Since the marbles sank, a given volume of marbles must weigh more than an equal volume of water. Thus the weight of the marbles afloat displaces more water than that displaced by their own volume when the marbles are actually in the water. Therefore, the water level in the tub will fall when the marbles are removed from the bowl and dropped into the tub.

ANSWERS

Whose Bananas?

Let n represent the number of bananas that each man received from the final division in the morning. Then there were 3n + 1 bananas left after the third man had taken his in the night.

We know he took a quantity equal to one more than half the quantity he left. Thus, when the third man came to take his bananas, there were 3/2(3n+1) + 1 remaining following the second man's take.

Similarly, there was 3/2 of this quantity plus one when the second man came for his and 3/2 of the resulting quantity plus one when the first man took his.

Thus, this initial quantity can be represented as

$$3/2[3/2(3/2[3N +1] + 1) +1] + 1 =$$

$$27/8(3N + 1) + 9/4 + 3/2 + 1 =$$

$$81/80 N + 27/80 + 9/4 + 3/2 + 1 = 10^{1/8} N + 8^{1/8}.$$

It is easily seen that 7 is the smallest whole number for n, which will give a whole number result. Thus, there were 79 bananas to start.

The next whole number value of n, which will give a whole number result is 15, but that would mean there was an initial quantity of 160.

Blank Page

www.ingramcontent.com/pod-product-compliance
Lightning Source LLC
Chambersburg PA
CBHW080501220526
45465CB00006B/2333